01	Pg. 10	Simple Beam — Uniformly Distributed Load	
02	Pg. 12	Simple Beam — Uniform Load Partially Distributed	
03	Pg. 14	Simple Beam — Uniform Load Partially Distributed at One End	
04	Pg. 16	Simple Beam — Uniform Load Partially Distributed at Each End	
05	Pg. 18	Simple Beam — Load Increasing Uniformly to One End	
06	Pg. 20	Simple Beam — Load Increasing Uniformly to Centre	
07	Pg. 22	Simple Beam — Concentrated Load at Centre	
08	Pg. 24	Simple Beam — Concentrated Load at Any Point	
09	Pg. 26	Simple Beam — Two Equal Concentrated Loads Symmetrically Placed	
10	Pg. 28	Simple Beam — Two Equal Concentrated Loads Unsymmetrically Placed	

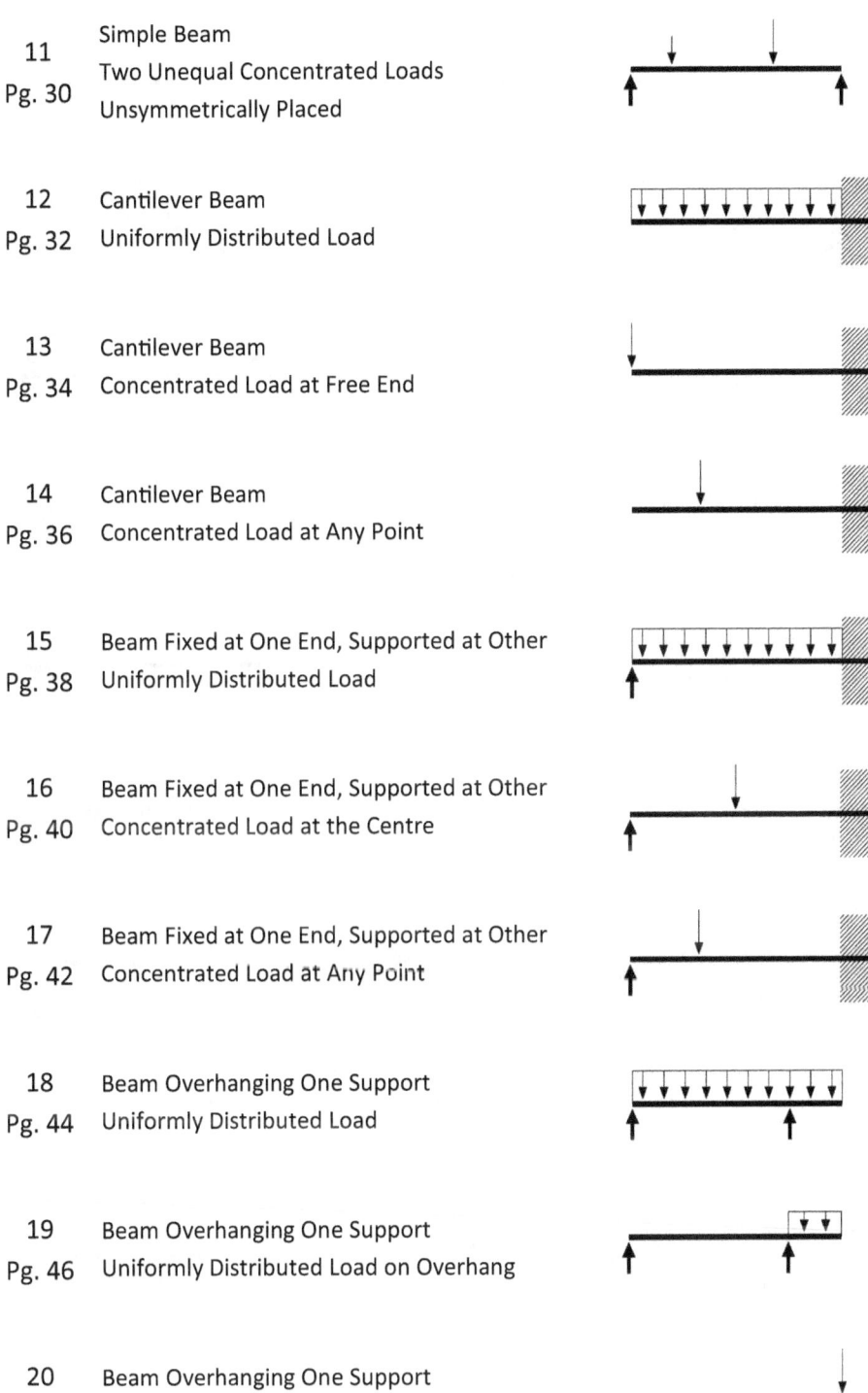

11 Pg. 30	Simple Beam Two Unequal Concentrated Loads Unsymmetrically Placed	
12 Pg. 32	Cantilever Beam Uniformly Distributed Load	
13 Pg. 34	Cantilever Beam Concentrated Load at Free End	
14 Pg. 36	Cantilever Beam Concentrated Load at Any Point	
15 Pg. 38	Beam Fixed at One End, Supported at Other Uniformly Distributed Load	
16 Pg. 40	Beam Fixed at One End, Supported at Other Concentrated Load at the Centre	
17 Pg. 42	Beam Fixed at One End, Supported at Other Concentrated Load at Any Point	
18 Pg. 44	Beam Overhanging One Support Uniformly Distributed Load	
19 Pg. 46	Beam Overhanging One Support Uniformly Distributed Load on Overhang	
20 Pg. 48	Beam Overhanging One Support Concentrated Load at End of overhang	

21 Pg. 50	Beam Overhanging One Support Concentrated Load at Any Point Between Supports	
22 Pg. 52	Beam Overhanging Both Supports Unequal Overhangs, Uniformly Distributed Load	
23 Pg. 54	Beam Fixed at Both Ends Uniformly Distributed Load	
24 Pg. 56	Beam Fixed at Both Ends Concentrated Load at Centre	
25 Pg. 58	Beam Fixed at Both Ends Concentrated Load at Any Point	
26 Pg. 60	Continuous Beam Two Equal Spans, Uniform Load on One Span	
27 Pg. 62	Continuous Beam Two Equal Spans, Concentrated Load at Centre of One Span	
28 Pg. 64	Continuous Beam Two Equal Spans, Concentrated Load at Any Point	
29 Pg. 66	Continuous Beam Two Equal Spans, Uniformly Distributed Load	
30 Pg. 68	Continuous Beam Two Equal Spans, Two Equal Concentrated Loads Symmetrically Placed	

#	Page	Description
31	Pg. 70	Continuous Beam — Two Unequal Spans, Uniformly Distributed Load
32	Pg. 72	Continuous Beam — Two Unequal Spans, Concentrated Load on Each Span Symmetrically Placed
33	Pg. 74	Continuous Beam — Three Span with Two Span Uniformly Distributed Load
34	Pg. 76	Continuous Beam — Three Span with End Span Uniformly Distributed Loads
35	Pg. 78	Continuous Beam — Three Span Beam with Uniformly Distributed Loads
36	Pg. 80	Continuous Beam — Four Span Beam with Unloaded Span
37	Pg. 82	Continuous Beam — Four Span with Two Spans Unloaded
38	Pg. 84	Continuous Beam — Four Span with UDL

This Reference Book

Presented within this book is a collection of equations used to analyse beams of different end conditions and loadings. The equations can be found separately in many engineering text books and reference documents. The purpose of this book is to bring the equations together into one convenient manuscript.

Essential for engineers.

Whilst every effort has been made to ensure the accuracy of the information presented within, the author and publishers accept no responsibility for any design produced from this publication.

Those using this publication assume all liability from its use.

Above or below

Typically bending moments are depicted above the centroid of a beam when positive and below when negative.

Shear Force can be presented either above or below the centroid provided that a clear indication of positive and negative is shown on the diagram.

Some institutions may prefer to reverse the conventions used in this publication however for simplicity, all diagrams have been produced with the positive above and negative below the centroid of the beam.

Symbols

Metric System (SI)

R = Reaction Load at Bearing Point (N)

V = Shear Force (N/m²)

M = Maximum Bending Moment (Nm)

ℓ = Span Length (m)

χ = Horizontal Distance from Reaction Point on Beam (m)

W = Total Uniform Load (N)

w = Load per Unit Length (N/m)

P = Total Concentrated Load (N)

Δ = Deflection (m)

E = Modulus of Elasticity (Pa)

I = Moment of Inertia (m⁴)

Symbols

Imperial System

R = Reaction Load at Bearing Point (lbs)

V = Shear Force (lbs)

M = Maximum Bending Moment (in lbs)

ℓ = Span Length (in)

χ = Horizontal Distance from Reaction Point on Beam (in)

W = Total Uniform Load (lbs)

w = Load per Unit Length (lbs/in)

P = Total Concentrated Load (lbs)

Δ = Deflection (in)

E = Modulus of Elasticity (psi)

I = Moment of Inertia (in^4)

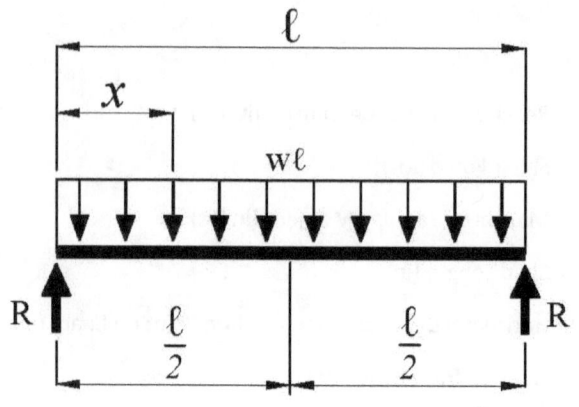

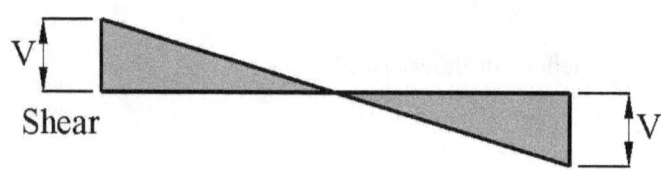
Shear

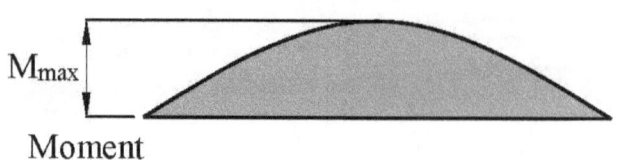
Moment

Simple Beam

Uniformly Distributed Load

$$R = V = \frac{w\ell}{2}$$

$$V_x = w\left(\frac{\ell}{2} - x\right)$$

$$M_{max} \text{ (at centre)} = \frac{w\ell^2}{8}$$

$$M_x = \frac{wx}{2}(\ell - x)$$

$$\Delta_{max} \text{ (at centre)} = \frac{5w\ell^4}{384\,EI}$$

$$\Delta_x = \frac{wx}{24\,EI}(\ell^3 - 2\ell x^2 + x^3)$$

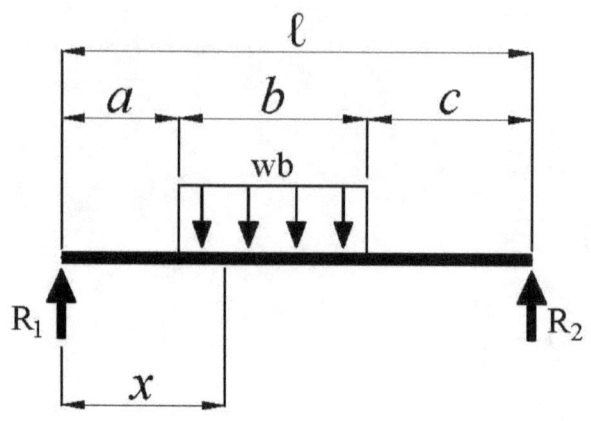

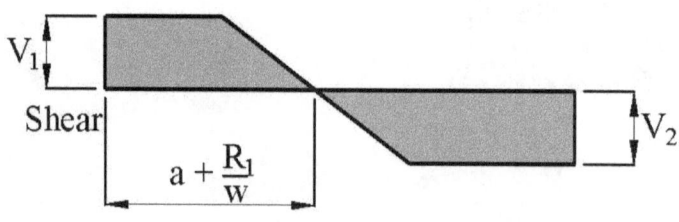

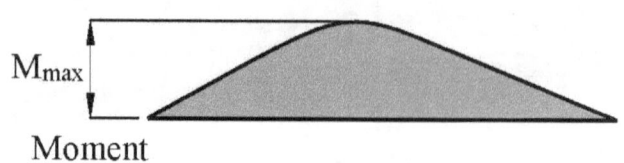

Simple Beam

Uniform Load Partially Distributed

$R_1 = V_1 \text{ (max when } a < c) = \dfrac{wb}{2\ell}(2c + b)$

$R_2 = V_2 \text{ (max when } a > c) = \dfrac{wb}{2\ell}(2a + b)$

$V_x \text{ (when } x > a \text{ \& } < (a+b)) = R_1 - w(x - a)$

$M_{max}\left(\text{at } x = a + \dfrac{R_1}{w}\right) = R_1\left(a + \dfrac{R_1}{2w}\right)$

$M_x \text{ (when } x < a) = R_1 x$

$M_x \text{ (when } x > a \text{ \& } < (a+b)) = R_1 x - \dfrac{w}{2}(x - a)^2$

$M_x \text{ (when } x > (a+b)) = R_2(\ell - x)$

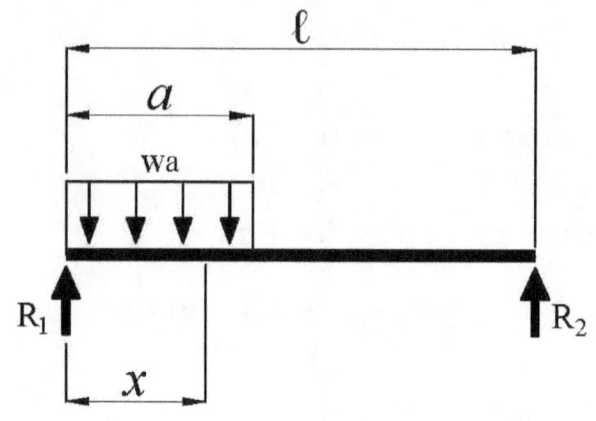

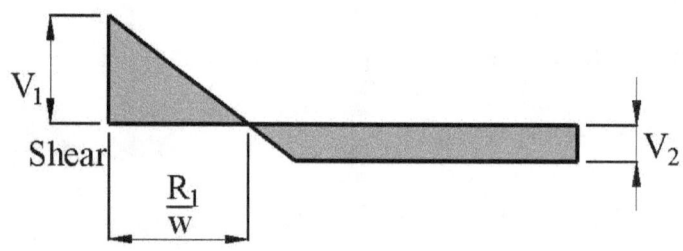

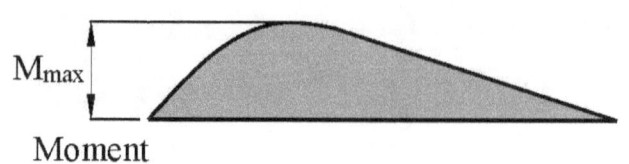

Simple Beam

Uniform Load Partially Distributed at One End

$$R_1 = V_1 = \frac{wa}{2\ell}(2\ell - a)$$

$$R_2 = V_2 = \frac{wa^2}{2\ell}$$

$$V_x \text{ (when } x < a) = R_1 - wx$$

$$M_{max}\left(at\ x = \frac{R_1}{w}\right) = \frac{R_1^2}{2w}$$

$$M_x \text{ (when } x < a) = R_1 x - \frac{wx^2}{2}$$

$$M_x \text{ (when } x > a) = R_2(\ell - x)$$

$$\Delta_x \text{ (when } x < a) = \frac{wx}{24\ EI\ell}(a^2(2\ell - a)^2 - 2ax^2(2\ell - a) + \ell x^3)$$

$$\Delta_x \text{ (when } x > a) = \frac{wa^2(\ell - x)}{24\ EI\ell}(4x\ell - 2x^2 - a^2)$$

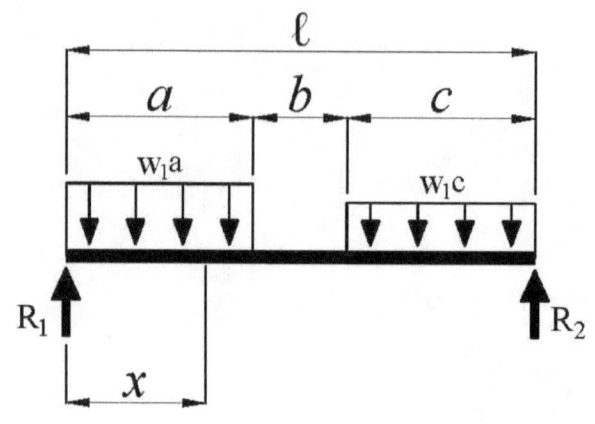

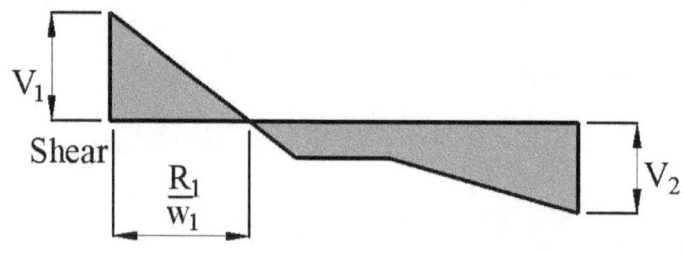

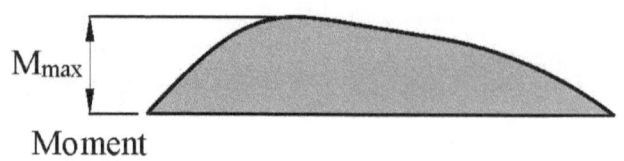

Simple Beam

Uniform Load Partially Distributed at Each End

$$R_1 = V_1 = \frac{w_1 a(2\ell - a) + w_2 c^2}{2\ell}$$

$$R_2 = V_2 = \frac{w_2 c(2\ell - c) + w_1 a^2}{2\ell}$$

V_x (when $x < a$) $= R_1 - w_1 x$

V_x (when $x > a$ & $< (a+b)$) $= R_1 - w_1 a$

V_x (when $x > (a+b)$) $= R_2 - w_2(\ell - x)$

$$M_{max} \left(at\ x = \frac{R_1}{w_1}\ when\ R_1 < w_1 a \right) = \frac{R_1^2}{2w_1}$$

$$M_{max} \left(at\ x = \ell - \frac{R_2}{w_2}\ when\ R_2 < w_2 c \right) = \frac{R_2^2}{2w_2}$$

M_x (when $x < a$) $= R_1 x - \dfrac{w_1 x^2}{2}$

M_x (when $x > a$ & $< (a+b)$) $= R_1 x - \dfrac{w_1 a}{2}(2x - a)$

M_x (when $x > (a+b)$) $= R_2(\ell - x) - \dfrac{w_2(\ell - x)^2}{2}$

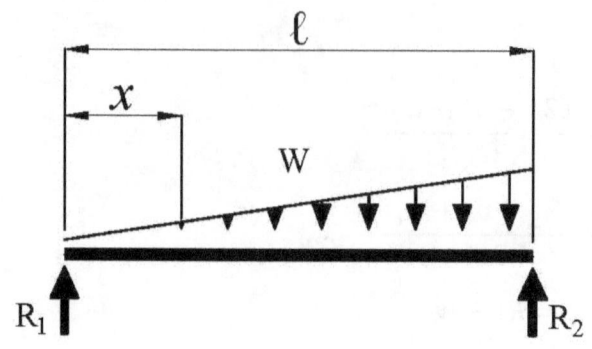

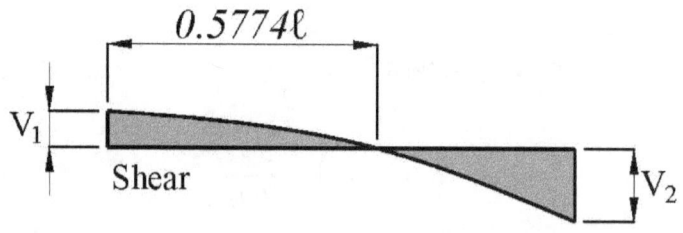

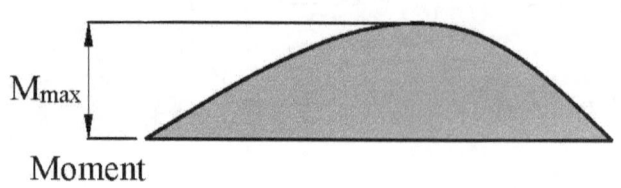

Simple Beam

Load Increasing Uniformly to One End

$R_1 = V_1 = \dfrac{W}{3}$

$R_2 = V_2 = \dfrac{2W}{3}$

$V_x = \dfrac{W}{3} - \dfrac{Wx^2}{\ell^2}$

$M_{max} \left(at\ x = \dfrac{\ell}{\sqrt{3}} = 0.5774\ell \right) = \dfrac{2W\ell}{9\sqrt{3}} = 0.1283W\ell$

$M_x = \dfrac{Wx}{3\ell^2}(\ell^2 - x^2)$

$\Delta_{max} \left(at\ x = \ell\sqrt{1 - \sqrt{\dfrac{8}{15}}} = 0.5193\ell \right) = 0.01304\dfrac{W\ell^3}{EI}$

$\Delta_x = \dfrac{Wx}{180\ EI\ell^2}(3x^4 - 10\ell^2 x^2 + 7\ell^4)$

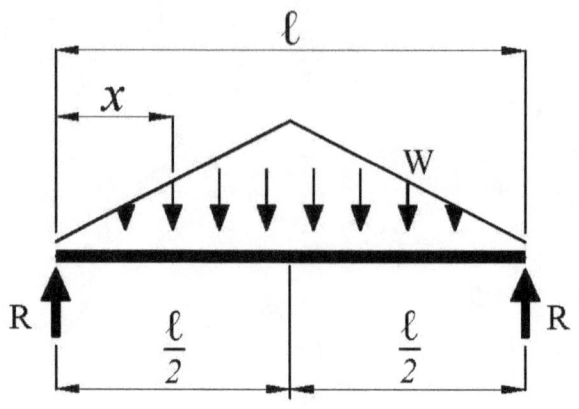

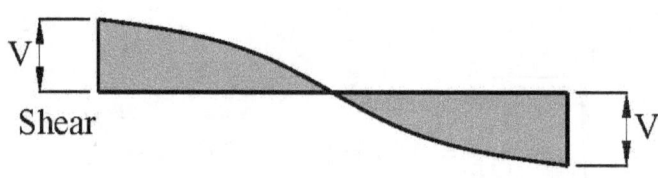
Shear

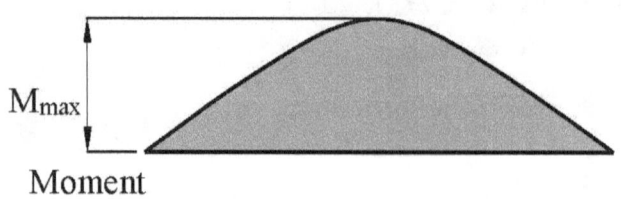
Moment

Simple Beam

Load Increasing Uniformly to Centre

$R = V = \dfrac{W}{2}$

$V_x \left(when\ x < \dfrac{\ell}{2}\right) = \dfrac{W}{2\ell^2}(\ell^2 - 4x^2)$

$M_{max}\ (at\ centre) = \dfrac{W\ell}{6}$

$M_x \left(when\ x < \dfrac{\ell}{2}\right) = Wx\left(\dfrac{1}{2} - \dfrac{2x^2}{3\ell^2}\right)$

$\Delta_{max}\ (at\ centre\ ) = \dfrac{W\ell^3}{60\ EI}$

$\Delta_x = \dfrac{Wx}{480\ EI\ell^2}(5\ell^2 - 4x^2)^2$

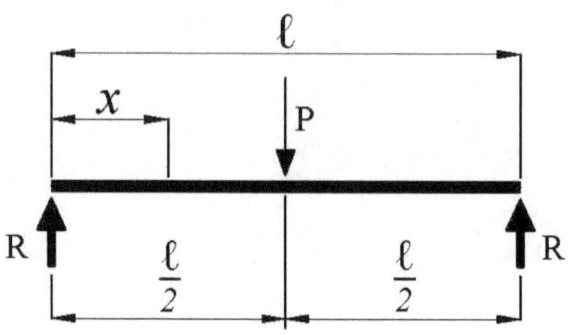

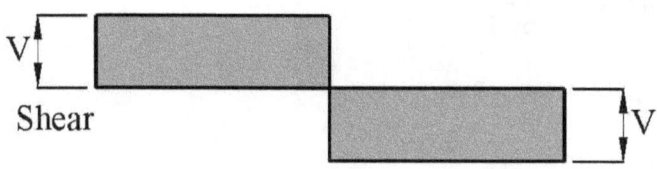
Shear

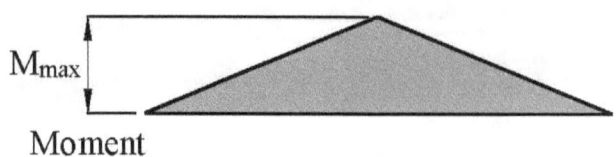
Moment

Simple Beam
Concentrated Load at Centre

$R = V = \dfrac{P}{2}$

$M_{max} \ (at \ point \ of \ load) = \dfrac{P\ell}{4}$

$M_x \left(when \ x < \dfrac{\ell}{2}\right) = \dfrac{Px}{2}$

$\Delta_{max} \ (at \ point \ of \ load \ ) = \dfrac{P\ell^3}{48 \ EI}$

$\Delta_x \left(when \ x < \dfrac{\ell}{2}\right) = \dfrac{Px}{48 \ EI}(3\ell^2 - 4x^2)$

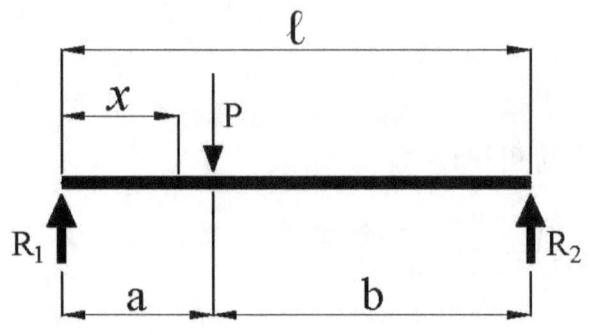

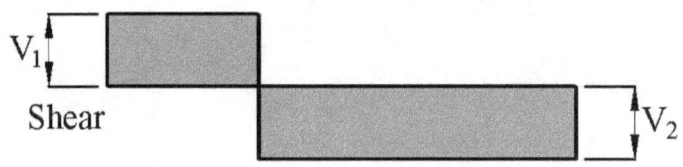

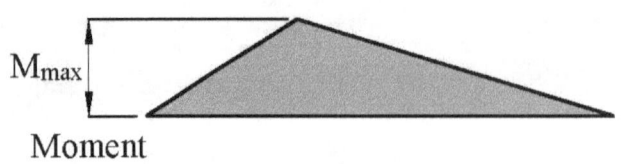

Simple Beam

Concentrated Load at Any Point

$R_1 = V_1 \ (max \ when \ a < b) = \dfrac{Pb}{\ell}$

$R_2 = V_2 \ (max \ when \ a > b) = \dfrac{Pa}{\ell}$

$M_{max} \ (at \ point \ of \ load) = \dfrac{Pab}{\ell}$

$M_x \ (when \ x < a) = \dfrac{Pbx}{\ell}$

$\Delta_{max} \left(at \ x = \sqrt{\dfrac{a(a+2b)}{3}} \ when \ a > b \right) = \dfrac{Pab(a+2b)\sqrt{3a(a+2b)}}{27 \ EI\ell}$

$\Delta_a \ (at \ point \ of \ load \ ) = \dfrac{Pa^2 b^2}{3 \ EI\ell}$

$\Delta_x \ (when \ x < a \ ) = \dfrac{Pbx}{6 \ EI\ell}(\ell^2 - b^2 - x^2)$

$\Delta_x \ (when \ x > a \ ) = \dfrac{Pa(\ell - x)}{6 \ EI\ell}(2\ell x - x^2 - a^2)$

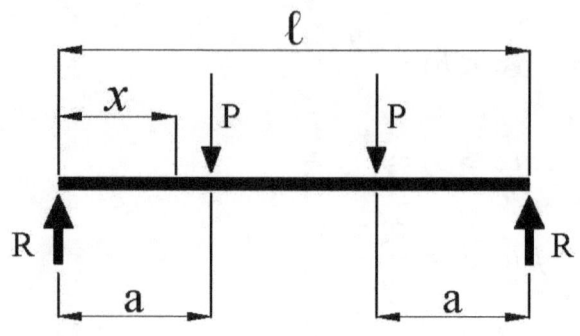

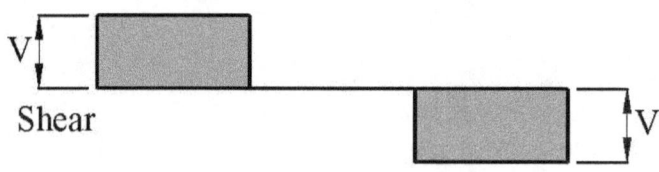

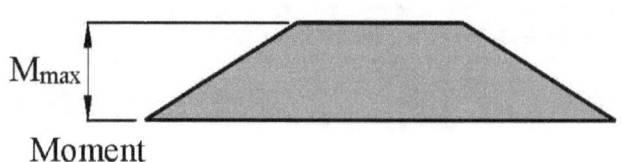

Simple Beam

Two Equal Concentrated Loads Symmetrically Placed

$R = V = P$

M_{max} (between loads) $= Pa$

M_x (when $x < a$) $= Px$

Δ_{max} (at centre) $= \dfrac{Pa}{24\,EI}(3\ell^2 - 4a^2)$

Δ_x (when $x < a$) $= \dfrac{Px}{6\,EI}(3\ell a - 3a^2 - x^2)$

Δ_x (when $x > a$ & $< (\ell - a)$) $= \dfrac{Pa}{6\,EI}(3\ell x - 3x^2 - a^2)$

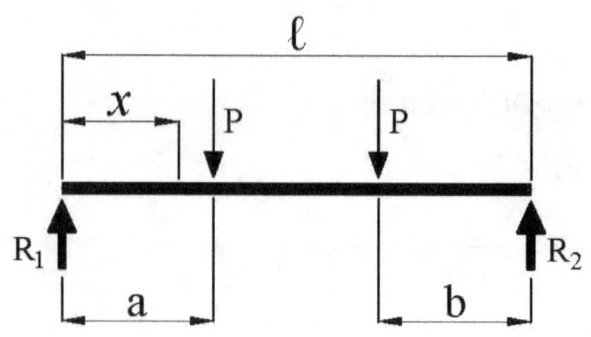

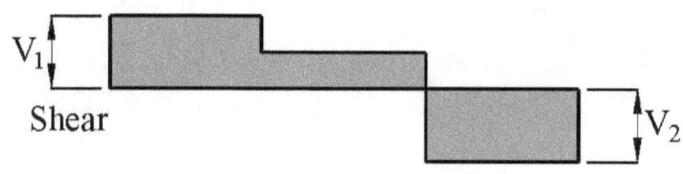

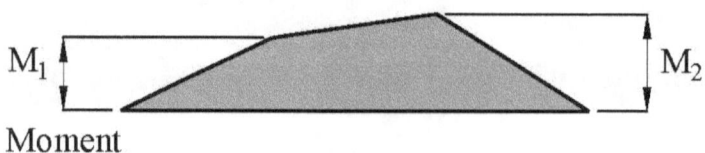

Simple Beam

Two Equal Concentrated Loads Unsymmetrically Placed

$R_1 = V_1 \ (max\ when\ a < b) = \dfrac{P}{\ell}(\ell - a + b)$

$R_2 = V_2 \ (max\ when\ a > b) = \dfrac{P}{\ell}(\ell - b + a)$

$V_x \ (when\ x > a\ \&\ < (\ell - b)) = \dfrac{P}{\ell}(b - a)$

$M_1 \ (max\ when\ a > b) = R_1 a$

$M_2 \ (max\ when\ a < b) = R_2 b$

$M_x \ (when\ x < a) = R_1 x$

$M_x \ (when\ x > a\ \&\ < (\ell - b)) = R_1 x - P(x - a)$

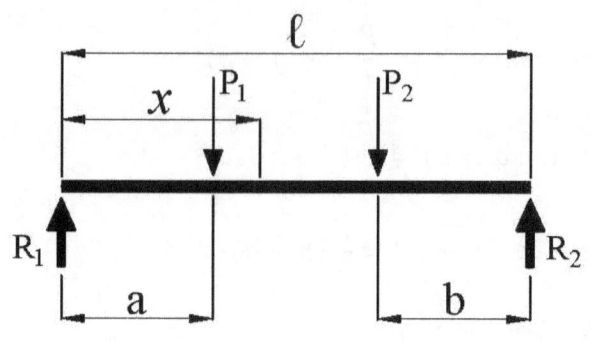

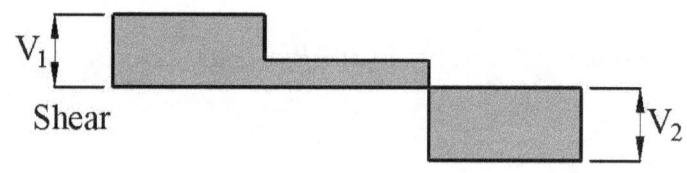

Shear

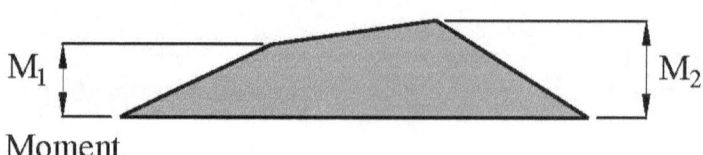

Moment

Simple Beam

Two Unequal Concentrated Loads Unsymmetrically Placed

$$R_1 = V_1 = \frac{P_1(\ell - a) + P_2 b}{\ell}$$

$$R_2 = V_2 = \frac{P_1 a + P_2(\ell - b)}{\ell}$$

$V_x \;(when\; x > a \;\&\; < (\ell - b)) = R_1 - P_1$

$M_1 \;(max\; when\; R_1 < P_1) = R_1 a$

$M_2 \;(max\; when\; R_2 < P_2) = R_2 b$

$M_x \;(when\; x < a) = R_1 x$

$M_x \;(when\; x > a \;\&\; < (\ell - b)) = R_1 x - P_1(x - a)$

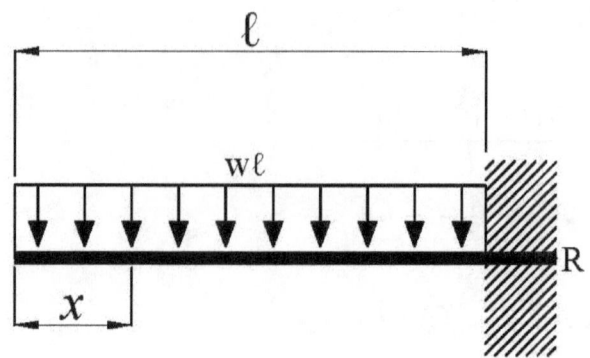

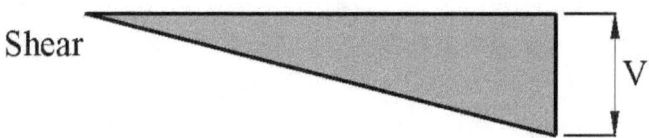

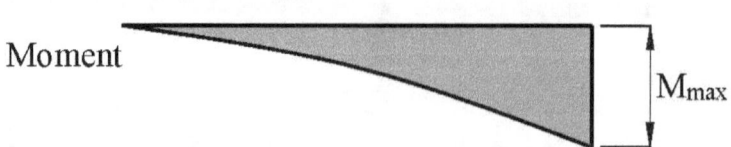

Cantilever Beam
Uniformly Distributed Load

$R = V = w\ell$

$V_x = wx$

$M_{max} \ (at\ fixed\ end) = \dfrac{w\ell^2}{2}$

$M_x = \dfrac{wx^2}{2}$

$\Delta_{max} \ (at\ free\ end) = \dfrac{w\ell^4}{8\ EI}$

$\Delta_x = \dfrac{w}{24\ EI}(x^4 - 4\ell^3 x + 3\ell^4)$

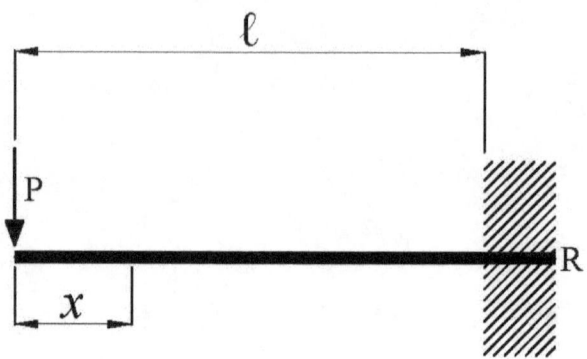

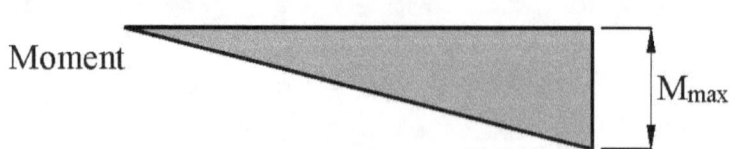

Cantilever Beam

Concentrated Load at Free End

$R = V = P$

$M_{max} \ (at \ fixed \ end) = P\ell$

$M_x = Px$

$\Delta_{max} \ (at \ free \ end) = \dfrac{P\ell^3}{3\,EI}$

$\Delta_x = \dfrac{P}{6\,EI}(2\ell^3 - 3\ell^2 x + x^3)$

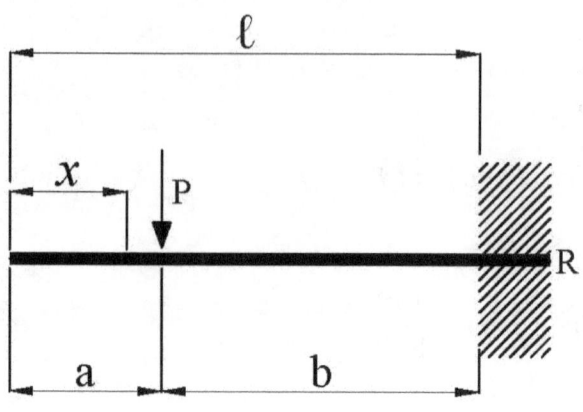

Shear

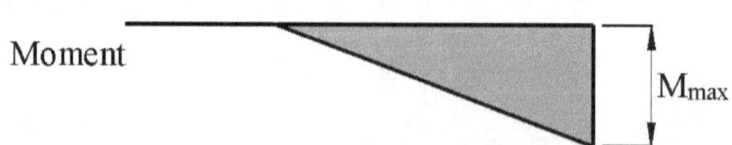

Moment

Cantilever Beam
Concentrated Load at Any Point

$R = V = P$

$M_{max}\ (at\ fixed\ end) = Pb$

$M_x\ (when\ x > a) = P(x - a)$

$\Delta_{max}\ (at\ free\ end) = \dfrac{Pb^2}{6\ EI}(3\ell - b)$

$\Delta_a\ (at\ point\ of\ load) = \dfrac{Pb^3}{3\ EI}$

$\Delta_x\ (when\ x < a) = \dfrac{Pb^2}{6\ EI}(3\ell - 3x - b)$

$\Delta_x\ (when\ x > a) = \dfrac{P(\ell - x)^2}{6\ EI}(3b - \ell + x)$

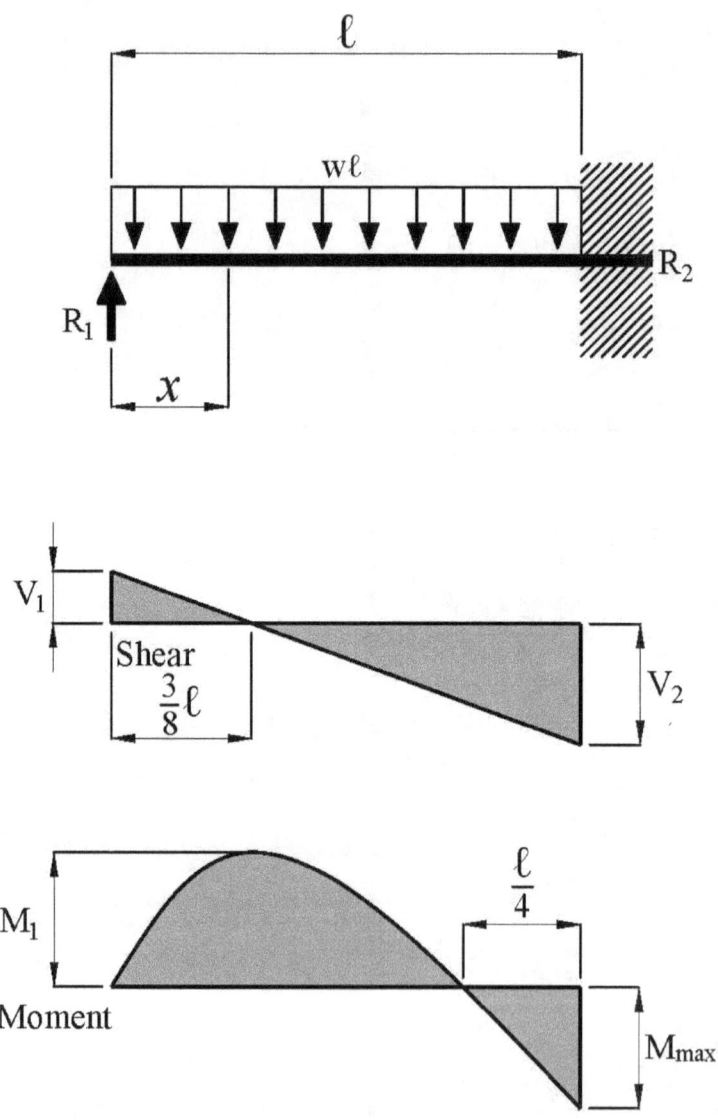

Beam Fixed at One End, Supported at Other

Uniformly Distributed Load

$R_1 = V_1 = \dfrac{3w\ell}{8}$

$R_2 = V_2 = \dfrac{5w\ell}{8}$

$V_x = R_1 - wx$

$M_{max} = \dfrac{w\ell^2}{8}$

$M_1 \left(at\ x = \dfrac{3}{8}\ell\right) = \dfrac{9}{128} w\ell^2$

$M_x = R_1 x - \dfrac{wx^2}{2}$

$\Delta_{max} \left(at\ x = \dfrac{\ell}{16}(1 + \sqrt{33}) = 0.4215\ell\right) = \dfrac{w\ell^4}{185\ EI}$

$\Delta_x = \dfrac{wx}{48\ EI}(\ell^3 - 3\ell x^2 + 2x^3)$

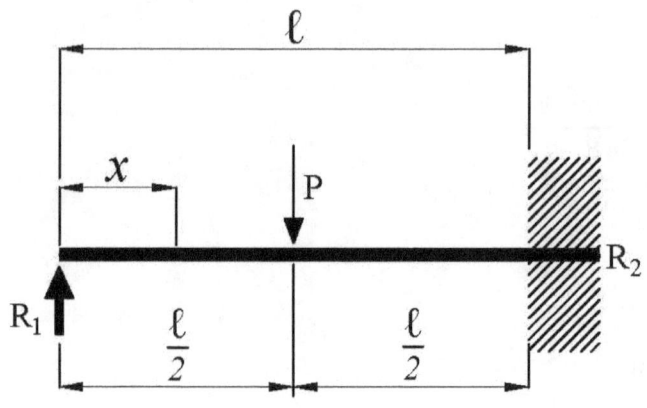
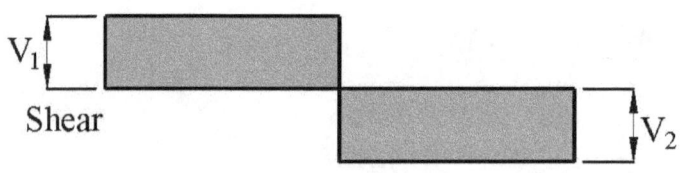
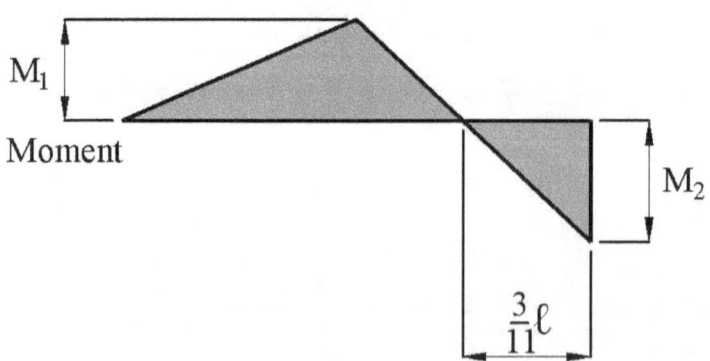

Beam Fixed at One End, Supported at Other

Concentrated Load at the Centre

$$R_1 = V_1 = \frac{5P}{16}$$

$$R_2 = V_2 = \frac{11P}{16}$$

$$M_{max} \text{ (at fixed end)} = \frac{3P\ell}{16}$$

$$M_1 \text{ (at point of load)} = \frac{5P\ell}{32}$$

$$M_x \left(\text{when } x < \frac{\ell}{2}\right) = \frac{5Px}{16}$$

$$M_x \left(\text{when } x > \frac{\ell}{2}\right) = P\left(\frac{\ell}{2} - \frac{11x}{16}\right)$$

$$\Delta_{max} \left(\text{at } x = \ell\sqrt{\frac{1}{5}} = 0.4472\ell\right) = \frac{P\ell^3}{48\,EI\sqrt{5}} = 0.009317 \frac{P\ell^3}{EI}$$

$$\Delta_x \text{ (at point of load)} = \frac{7P\ell^3}{768\,EI}$$

$$\Delta_x \left(\text{when } x < \frac{\ell}{2}\right) = \frac{Px}{96\,EI}(3\ell^2 - 5x^2)$$

$$\Delta_x \left(\text{when } x > \frac{\ell}{2}\right) = \frac{P}{96\,EI}(x - \ell)^2(11x - 2\ell)$$

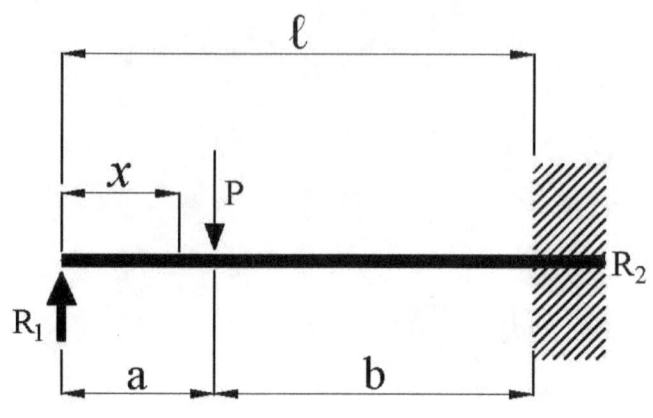

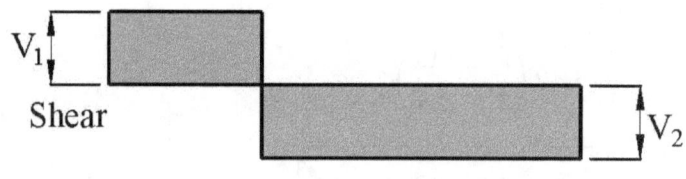

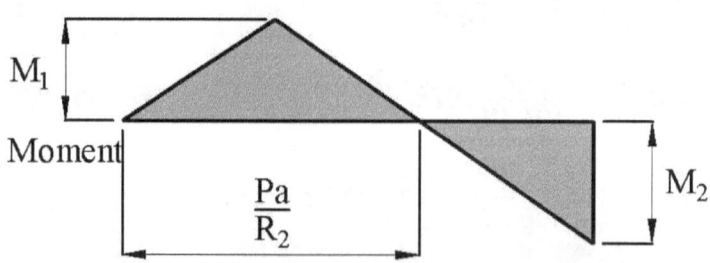

Beam Fixed at One End, Supported at Other

Concentrated Load at Any Point

$$R_1 = V_1 = \frac{Pb^2}{2\ell^3}(a + 2\ell)$$

$$R_2 = V_2 = \frac{Pa}{2\ell^3}(3\ell^2 - a^2)$$

M_1 (at point of load) $= R_1 a$

$$M_2 \text{ (at fixed end)} = \frac{Pab}{2\ell^2}(a + \ell)$$

M_x (when $x < a$) $= R_1 x$

M_x (when $x > a$) $= R_1 x - P(x - a)$

$$\Delta_{max}\left(when\ a < 0.414\ell\ at\ x = \ell\frac{\ell^2 + a^2}{3\ell^2 - a^2}\right) = \frac{Pa}{3\ EI}\frac{(\ell^2 - a^2)^3}{(3\ell^2 - a^2)^2}$$

$$\Delta_{max}\left(when\ a > 0.414\ell\ at\ x = \ell\sqrt{\frac{a}{2\ell + a}}\right) = \frac{Pab^2}{6\ EI}\sqrt{\frac{a}{2\ell + a}}$$

$$\Delta_a \text{ (at point of load)} = \frac{Pa^2 b^3}{12\ EI\ell^3}(3\ell + a)$$

$$\Delta_x \text{ (when } x < a) = \frac{Pb^2 x}{12\ EI\ell^3}(3a\ell^2 - 2\ell x^2 - ax^2)$$

$$\Delta_x \text{ (when } x > a) = \frac{Pa}{12\ EI\ell^3}(\ell - x)^2(3\ell^2 x - a^2 x - 2a^2 \ell)$$

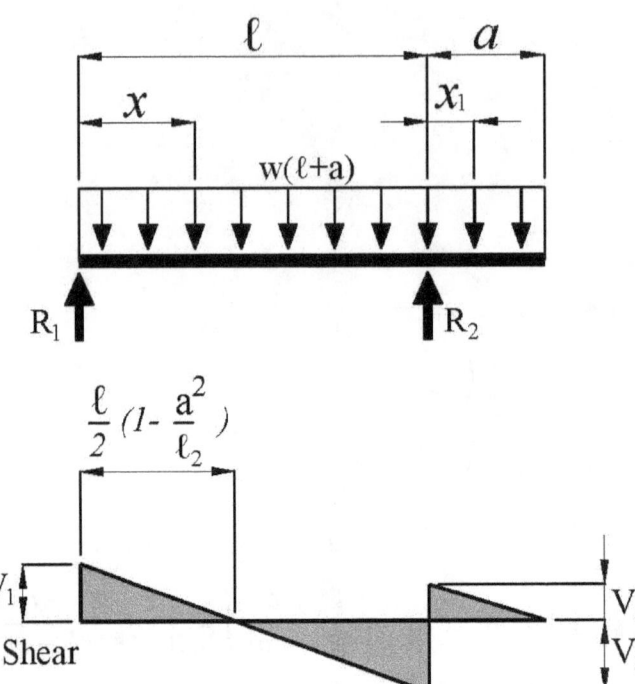
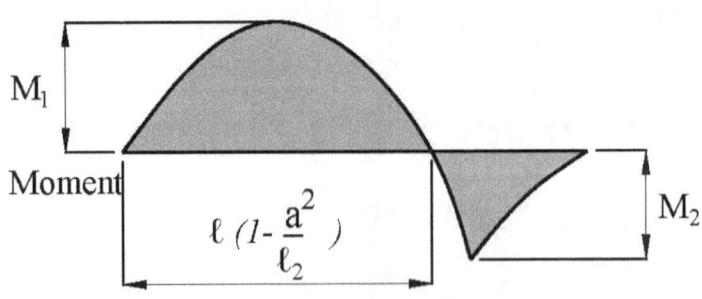

Beam Overhanging One Support

Uniformly Distributed Load

$R_1 = V_1 = \dfrac{w}{2\ell}(\ell^2 - a^2)$

$R_2 = V_2 + V_3 = \dfrac{w}{2\ell}(\ell + a)^2$

$V_2 = wa$

$V_3 = \dfrac{w}{2\ell}(\ell^2 + a^2)$

$V_x \text{ (between supports)} = R_1 - wx$

$V_{x1} \text{ (For Overhang)} = w(a - x_1)$

$M_1 \left(at\ x = \dfrac{\ell}{2}\left[1 - \dfrac{a^2}{\ell^2}\right] \right) = \dfrac{w}{8\,\ell^2}(\ell + a)^2(\ell - a)^2$

$M_2 \text{ (at } R_2\text{)} = \dfrac{wa^2}{2}$

$M_x \text{ (between supports)} = \dfrac{wx}{2\ell}(\ell^2 - a^2 - x\ell)$

$M_{x1} \text{ (for overhang)} = \dfrac{w}{2}(a - x_1)^2$

$\Delta_x \text{ (between supports)} = \dfrac{wx}{24\,EI\ell}(\ell^4 - 2\ell^2 x^2 + \ell x^3 - 2a^2\ell^2 + 2a^2 x^2)$

$\Delta_{x1} \text{ (for overhang)} = \dfrac{wx_1}{24\,EI}(4a^2\ell - \ell^3 + 6a^2 x_1 - 4a x_1^2 + x_1^3)$

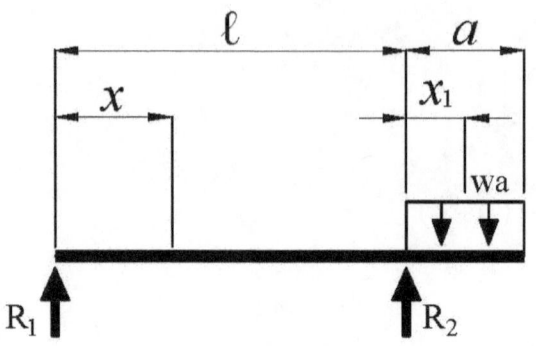

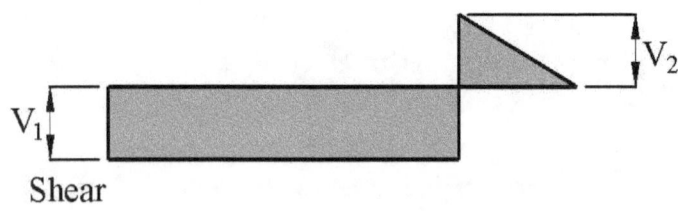

Shear

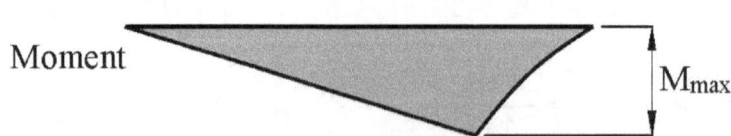

Moment

Beam Overhanging One Support

Uniformly Distributed Load on Overhang

$R_1 = V_1 = \dfrac{wa^2}{2\ell}$

$R_2 = V_1 + V_2 = \dfrac{wa}{2\ell}(2\ell + a)$

$V_2 = wa$

$V_{x1}\ (For\ Overhang) = w(a - x_1)$

$M_{max}\ (at\ R_2) = \dfrac{wa^2}{2}$

$M_x\ (between\ supports) = \dfrac{wa^2 x}{2\ell}$

$M_{x1}\ (for\ overhang) = \dfrac{w}{2}(a - x_1)^2$

$\Delta_{max}\left(between\ supports\ at\ x = \dfrac{\ell}{\sqrt{3}}\right) = \dfrac{wa^2\ell^2}{18\sqrt{3}\,EI} = 0.03208\dfrac{wa^2\ell^2}{EI}$

$\Delta_{max}\ (for\ overhang\ at\ x_1 = a) = \dfrac{wa^3}{24\,EI}(4\ell + 3a)$

$\Delta_x\ (between\ supports) = \dfrac{wa^2 x}{12\,EI\ell}(\ell^2 - x^2)$

$\Delta_{x1}\ (for\ overhang) = \dfrac{wx_1}{24\,EI}(4a^2\ell + 6a^2 x_1 - 4a x_1^2 + x_1^3)$

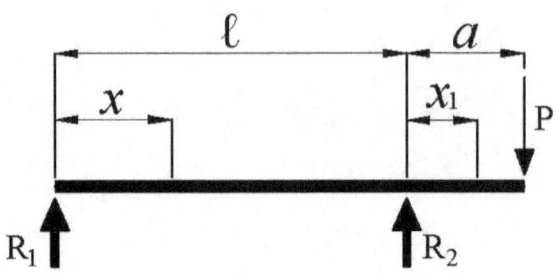

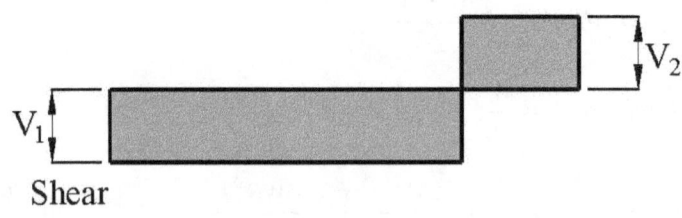

Shear

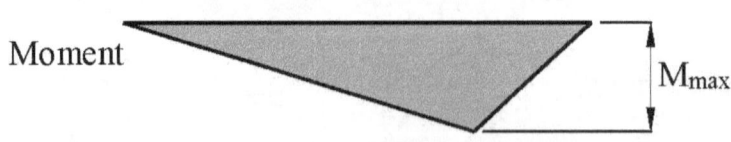

Moment

48

Beam Overhanging One Support
Concentrated Load at End of overhang

$R_1 = V_1 = \dfrac{Pa}{\ell}$

$R_2 = V_1 + V_2 = \dfrac{P}{\ell}(\ell + a)$

$V_2 = P$

M_{max} (at R_2) $= Pa$

M_x (between supports) $= \dfrac{Pax}{\ell}$

M_{x1} (for overhang) $= P(a - x_1)$

Δ_{max} (between supports at $x = \dfrac{\ell}{\sqrt{3}}$) $= \dfrac{Pa\ell^2}{9\sqrt{3}\,EI} = 0.06415\dfrac{Pa\ell^2}{EI}$

Δ_{max} (for overhang at $x_1 = a$) $= \dfrac{Pa^2}{3\,EI}(\ell + a)$

Δ_x (between supports) $= \dfrac{Pax}{6\,EI\ell}(\ell^2 - x^2)$

Δ_x (for overhang) $= \dfrac{Px_1}{6\,EI}(2a\ell + 3ax_1 - x_1^2)$

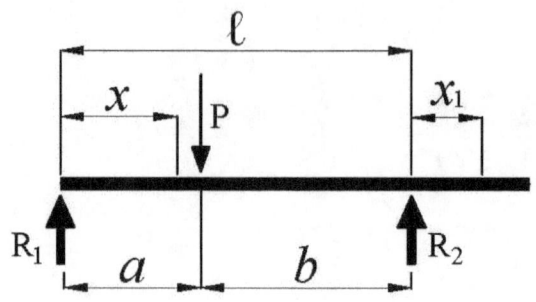

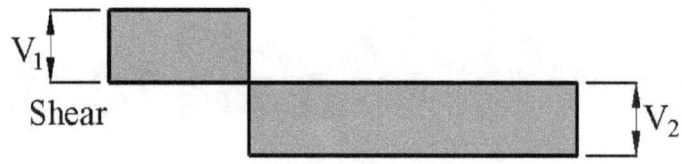

Shear

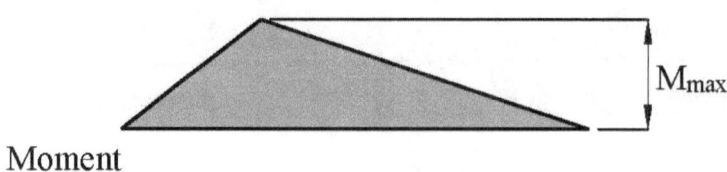

Moment

Beam Overhanging One Support

Concentrated Load at Any Point Between Supports

$$R_1 = V_1 \ (max \ when \ a < b) = \frac{Pb}{\ell}$$

$$R_2 = V_2 \ (max \ when \ a > b) = \frac{Pa}{\ell}$$

$$M_{max} \ (at \ point \ of \ load) = \frac{Pab}{\ell}$$

$$M_x \ (when \ x < a) = \frac{Pbx}{\ell}$$

$$\Delta_{max} \left(at \ x = \sqrt{\frac{a(a+2b)}{3}} \ when \ a > b \right) = \frac{Pab(a+2b)\sqrt{3a(a+2b)}}{27 \ EI\ell}$$

$$\Delta_a \ (at \ point \ of \ load) = \frac{Pa^2 b^2}{3 \ EI\ell}$$

$$\Delta_x \ (when \ x < a) = \frac{Pbx}{6 \ EI\ell}(\ell^2 - b^2 - x^2)$$

$$\Delta_x \ (when > a) = \frac{Pa(\ell - x)}{6 \ EI\ell}(2\ell x - x^2 - a^2)$$

$$\Delta_{x1} = \frac{Pabx_1}{6 \ EI\ell}(\ell + a)$$

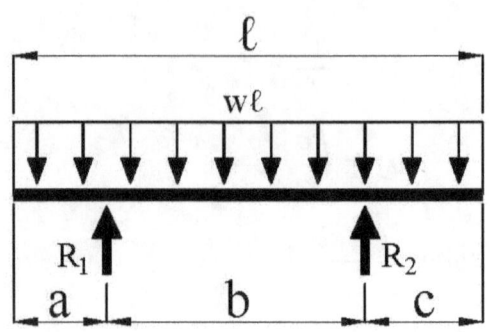

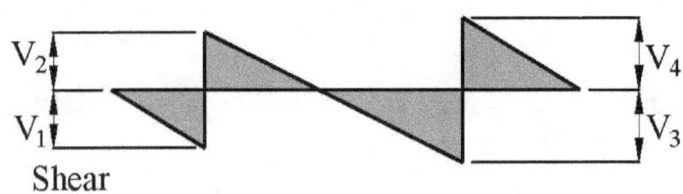

Shear

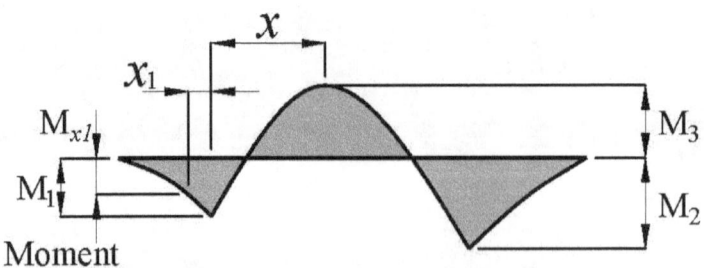

Moment

Beam Overhanging Both Supports
Unequal Overhangs, Uniformly Distributed Load

$R_1 = \dfrac{w\ell(\ell - 2c)}{2b}$

$R_2 = \dfrac{w\ell(\ell - 2a)}{2b}$

$V_1 = wa$

$V_2 = R_1 - V_1$

$V_3 = R_2 - V_4$

$V_4 = wc$

$V_{x_1} = V_1 - wx_1$

$V_x \text{ (when } x < b) = R_1 - w(a + x)$

$V_{max} \text{ (when } a < c) = R_2 - wc$

$M_1 = -\dfrac{wa^2}{2}$

$M_2 = -\dfrac{wc^2}{2}$

$M_3 = R_1 \left(\dfrac{R_1}{2w} - a\right)$

$M_x \left(max \text{ when } x = \dfrac{R_1}{w} - a\right) = R_1 x - \dfrac{w(a + x)^2}{2}$

53

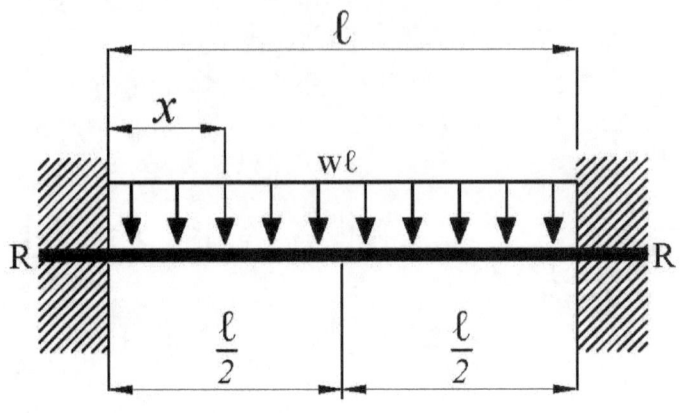

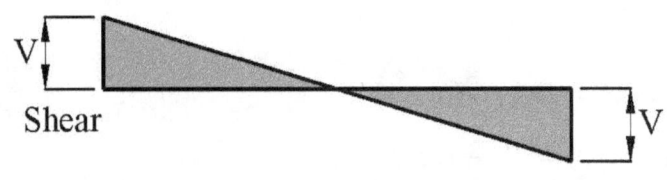

Shear

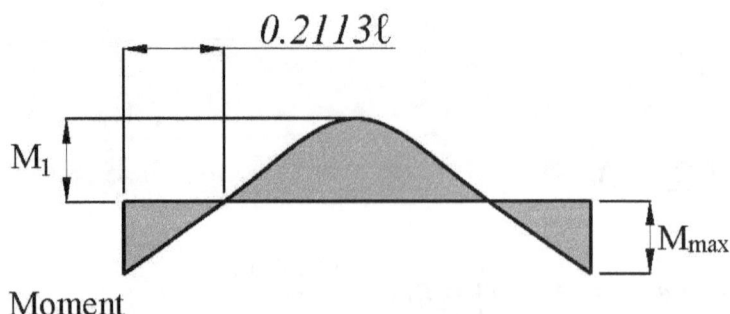

Moment

Beam Fixed at Both Ends

Uniformly Distributed Load

$$R = V = \frac{w\ell}{2}$$

$$V_x = w\left(\frac{\ell}{2} - x\right)$$

$$M_{max}(at\ ends) = \frac{w\ell^2}{12}$$

$$M_1\ (at\ centre) = \frac{w\ell^2}{24}$$

$$M_x = \frac{w}{12}(6\ell x - \ell^2 - 6x^2)$$

$$\Delta_{max}\ (at\ centre) = \frac{w\ell^4}{384\ EI}$$

$$\Delta_x = \frac{wx^2}{24\ EI}(\ell - x)^2$$

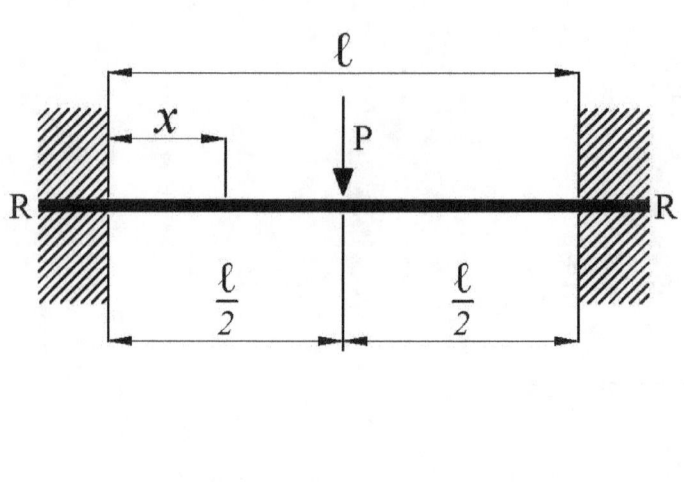
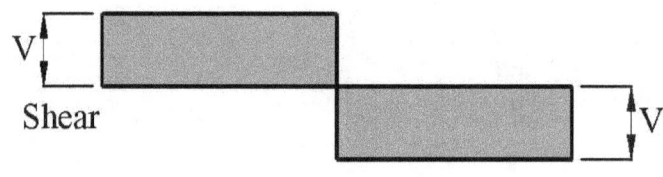
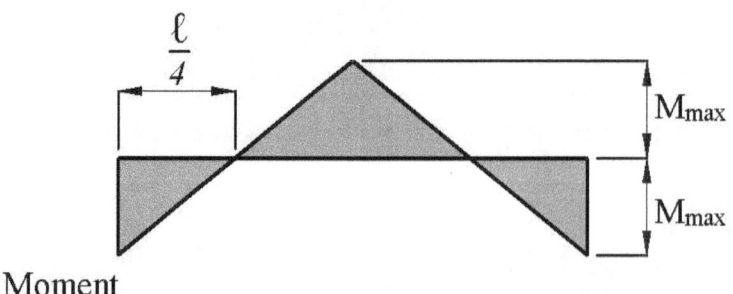

Beam Fixed at Both Ends

Concentrated Load at Centre

$R = V = \dfrac{P}{2}$

$M_{max} \ (at \ centre \ \& \ ends) = \dfrac{P\ell}{8}$

$M_x \left(when \ x < \dfrac{\ell}{2} \right) = \dfrac{P}{8}(4x - \ell)$

$\Delta_{max} \ (at \ centre) = \dfrac{P\ell^3}{192 \ EI}$

$\Delta_x \left(when \ x < \dfrac{\ell}{2} \right) = \dfrac{Px^2}{48 \ EI}(3\ell - 4x)$

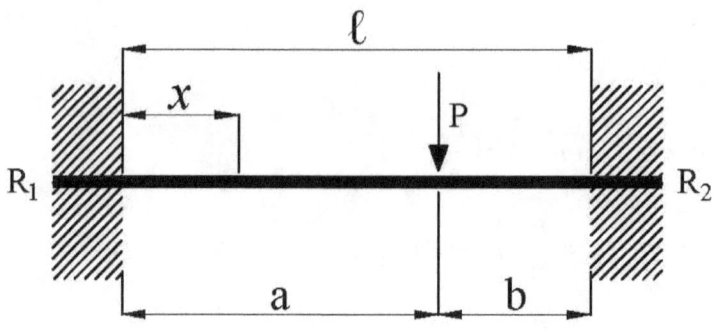

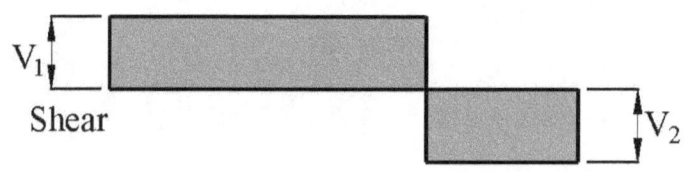

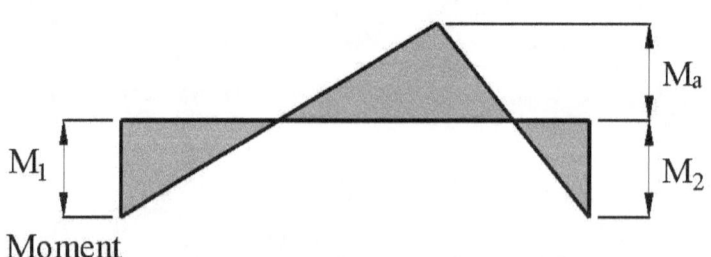

Beam Fixed at Both Ends

Concentrated Load at Any Point

$R_1 = V_1 \ (max\ when\ a < b) = \dfrac{Pb^2}{\ell^3}(3a + b)$

$R_2 = V_2 \ (max\ when\ a > b) = \dfrac{Pa^2}{\ell^3}(a + 3b)$

$M_1 \ (max\ when\ a < b) = \dfrac{Pab^2}{\ell^2}$

$M_2 \ (max\ when\ a > b) = \dfrac{Pa^2 b}{\ell^2}$

$M_a \ (at\ point\ of\ load) = \dfrac{2Pa^2 b^2}{\ell^3}$

$M_x = (when\ x < a) = R_1 x - \dfrac{Pab^2}{\ell^2}$

$\Delta_{max} \left(when\ a > b\ at\ x = \dfrac{2a\ell}{3a + b}\right) = \dfrac{2Pa^3 b^2}{3\ EI(3a + b)^2}$

$\Delta_a \ (at\ point\ of\ load) = \dfrac{Pa^3 b^3}{3\ EI\ell^3}$

$\Delta_x \ (when\ x < a) = \dfrac{Pb^2 x^2}{6\ EI\ell^3}(3a\ell - 3ax - bx)$

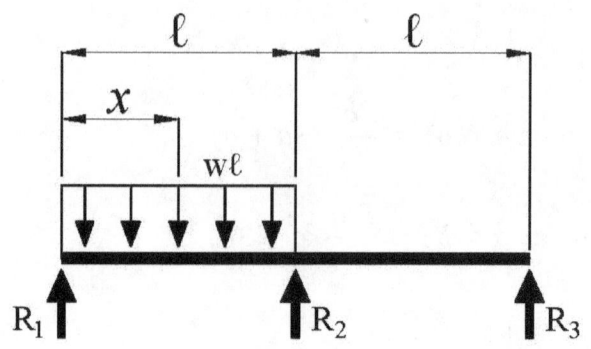
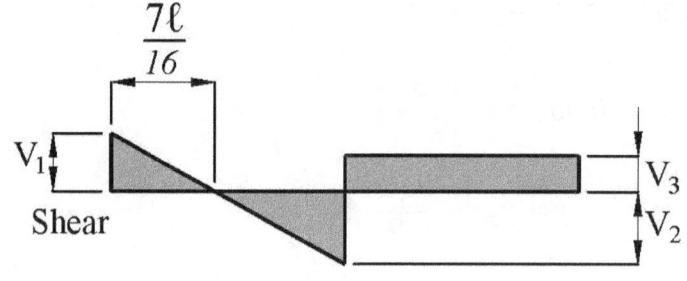
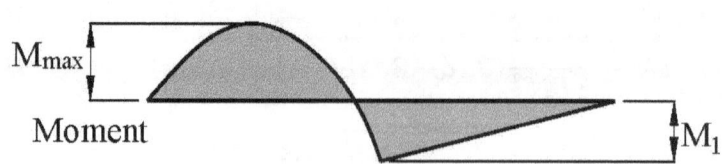

Continuous Beam

Two Equal Spans, Uniform Load on One Span

$R_1 = V_1 = \dfrac{7}{16} w\ell$

$R_2 = V_2 + V_3 = \dfrac{5}{8} w\ell$

$R_3 = V_3 = -\dfrac{1}{16} w\ell$

$V_2 = \dfrac{9}{16} w\ell$

$M_{max} \left(at\ x = \dfrac{7}{16}\ell \right) = \dfrac{49}{512} w\ell^2$

$M_1\ (at\ support\ R_2) = \dfrac{1}{16} w\ell^2$

$M_x\ (when\ x < \ell) = \dfrac{wx}{16}(7\ell - 8x)$

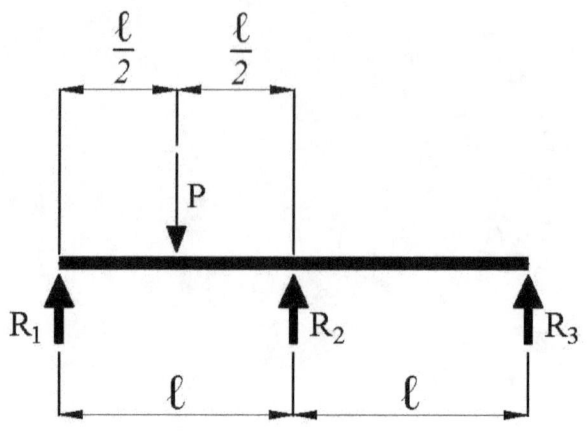

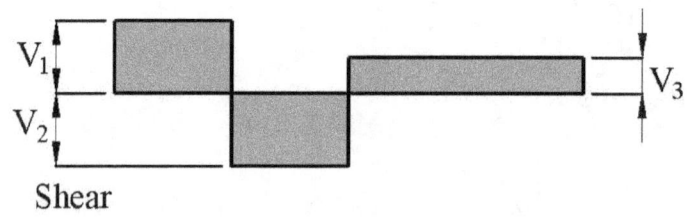

Shear

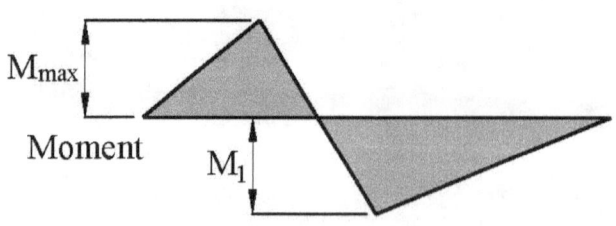

Moment

Continuous Beam

Two Equal Spans, Concentrated Load at Centre of One Span

$R_1 = V_1 = \dfrac{13}{32}P$

$R_2 = V_2 + V_3 = \dfrac{11}{16}P$

$R_3 = V_3 = -\dfrac{3}{32}P$

$V_2 = \dfrac{19}{32}P$

$M_{max}\ (at\ point\ of\ load) = \dfrac{13}{64}P\ell$

$M_1\ (at\ support\ R_2) = \dfrac{3}{32}P\ell$

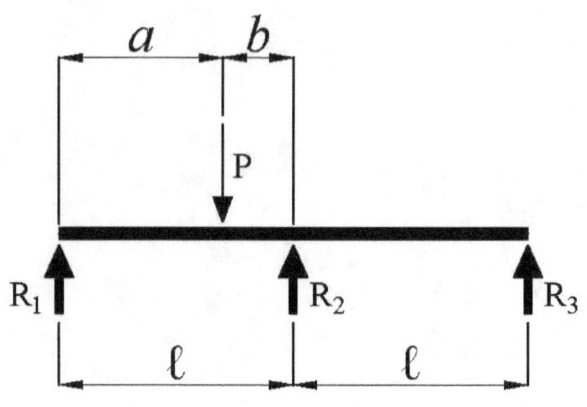

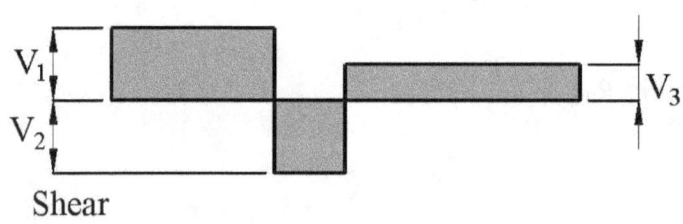

Shear

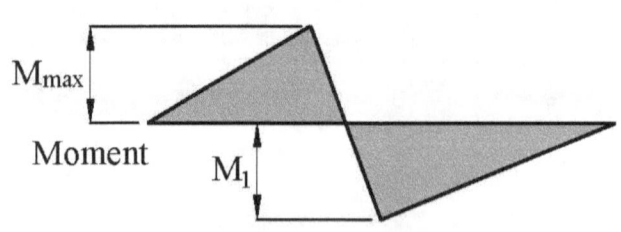

Moment

64

Continuous Beam

Two Equal Spans, Concentrated Load at Any Point

$$R_1 = V_1 = \frac{Pb}{4\ell^3}\left(4\ell^2 - a(\ell + a)\right)$$

$$R_2 = V_2 + V_3 = \frac{Pa}{2\ell^3}\left(2\ell^2 + b(\ell + a)\right)$$

$$R_3 = V_3 = -\frac{Pab}{4\ell^3}(\ell + a)$$

$$V_2 = \frac{Pa}{4\ell^3}\left(4\ell^2 + b(\ell + a)\right)$$

$$M_{max} \text{ (at point of load)} = \frac{Pab}{4\ell^3}\left(4\ell^2 - a(\ell + a)\right)$$

$$M_1 \text{ (at Support } R_2) = \frac{Pab}{4\ell^2}(\ell + a)$$

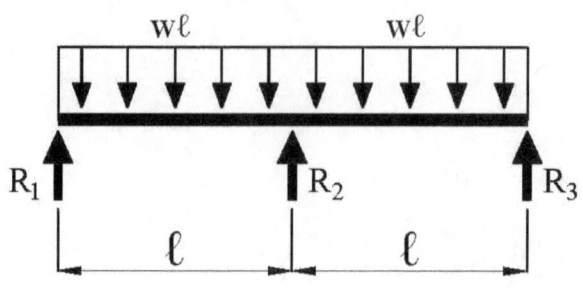

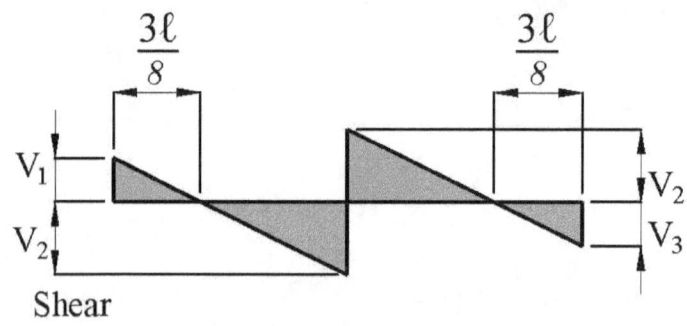

Shear

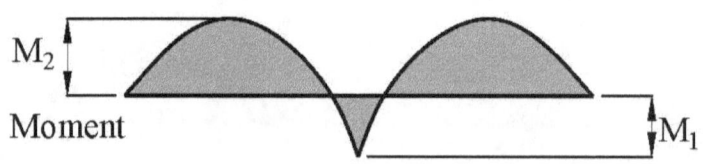

Moment

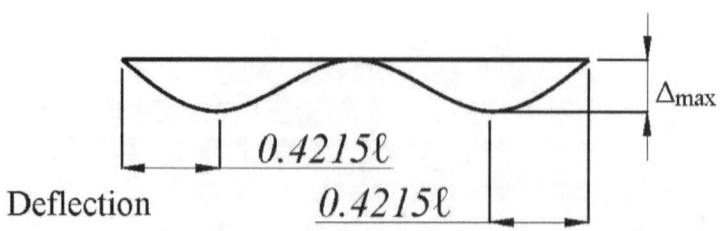

Deflection

Continuous Beam
Two Equal Spans, Uniformly Distributed Load

$$R_1 = V_1 = R_3 = V_3 = \frac{3w\ell}{8}$$

$$R_2 = \frac{10w\ell}{8}$$

$$V_2 = V_{max} = \frac{5w\ell}{8}$$

$$M_1 = \frac{w\ell^2}{8}$$

$$M_2 \left(at\ \frac{3\ell}{8} \right) = \frac{9w\ell^2}{128}$$

$$\Delta_{max}\ (at\ 0.4215\ell\ approx\ from\ R_1 and\ R_3) = \frac{w\ell^4}{185\ EI}$$

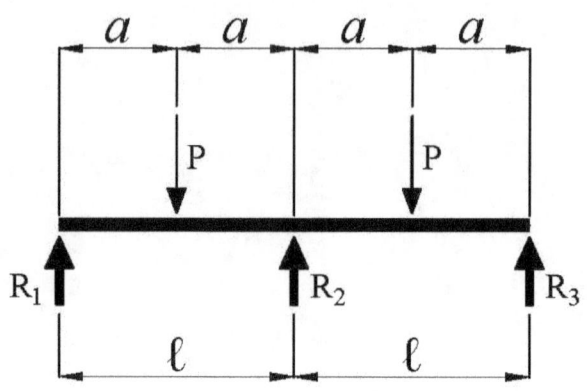
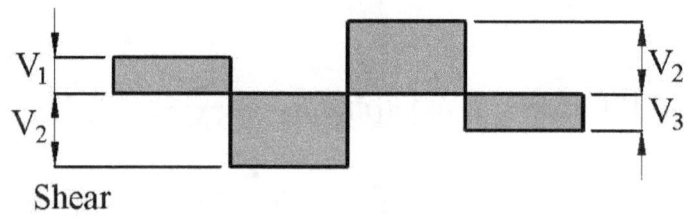

Shear

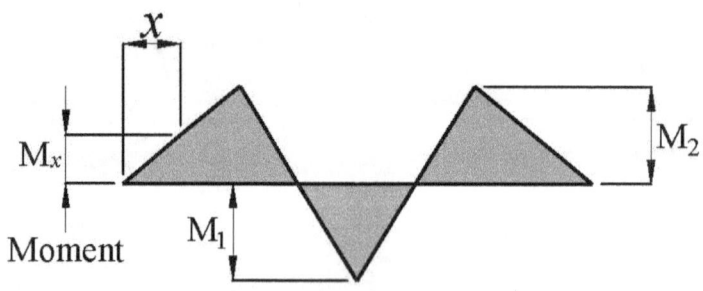

Moment

Continuous Beam

Two Equal Spans, Two Equal Concentrated Loads Symmetrically Placed

$$R_1 = V_1 = R_3 = V_3 = \frac{5P}{16}$$

$$R_2 = 2V_2 = \frac{11P}{8}$$

$$V_2 = P - R_1 = \frac{11P}{16}$$

$$V_{max} = V_2$$

$$M_1 = -\frac{3P\ell}{16}$$

$$M_2 = \frac{5P\ell}{32}$$

$$M_x \text{ (when } x < a) = R_1 x$$

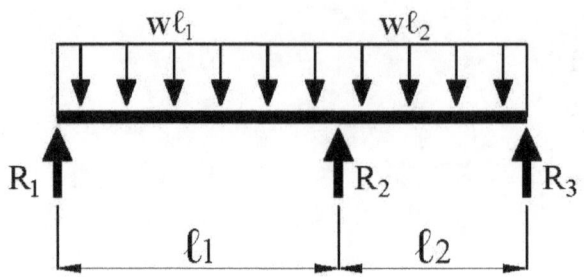

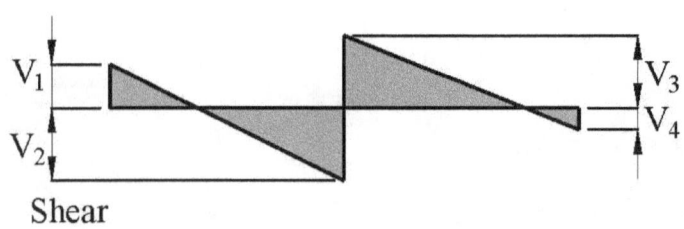

Shear

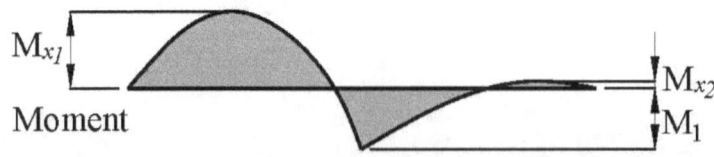

Moment

Continuous Beam

Two Unequal Spans, Uniformly Distributed Load

$$R_1 = \frac{M_1}{\ell_1} + \frac{w\ell_1}{2}$$

$$R_2 = w\ell_1 + w\ell_2 - R_1 - R_3$$

$$R_3 = V_4 = \frac{M_1}{\ell_2} + \frac{w\ell_2}{2}$$

$$V_1 = R_1$$

$$V_2 = w\ell_1 - R_1$$

$$V_3 = w\ell_2 - R_3$$

$$V_4 = R_3$$

$$M_1 = -\frac{w\ell_2^3 + w\ell_1^3}{8(\ell_1 + \ell_2)}$$

$$M_{x1}\left(when\ x_1 = \frac{R_1}{w}\right) = R_1 x_1 - \frac{wx_1^2}{2}$$

$$M_{x2}\left(when\ x_2 = \frac{R_3}{w}\right) = R_3 x_2 - \frac{wx_2^2}{2}$$

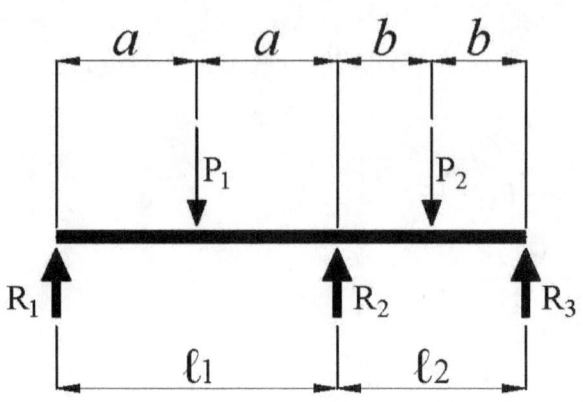

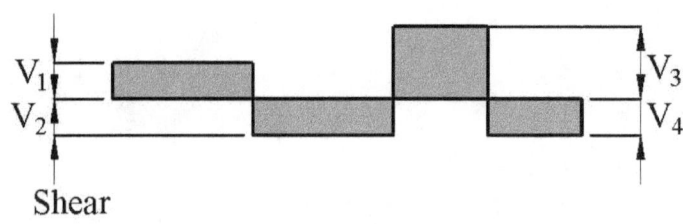

Shear

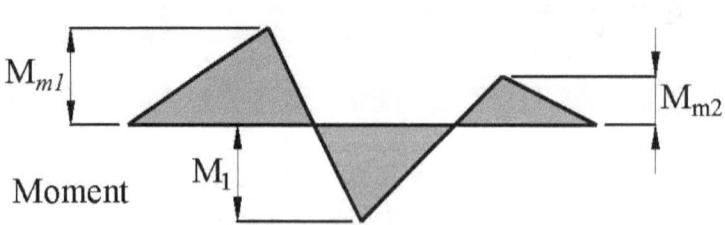

Moment

Continuous Beam
Two Unequal Spans, Concentrated Load on Each Span Symmetrically Placed

$$R_1 = \frac{M_1}{\ell_1} + \frac{P_1}{2}$$

$$R_2 = P_1 + P_2 - R_1 - R_3$$

$$R_3 = \frac{M_1}{\ell_2} + \frac{P_2}{2}$$

$$V_1 = R_1$$

$$V_2 = P_1 - R_1$$

$$V_3 = P_2 - R_3$$

$$V_4 = R_3$$

$$M_1 = -\frac{3}{16}\left(\frac{P_1\ell_1^2 + P_2\ell_2^2}{\ell_1 + \ell_2}\right)$$

$$M_{m1} = R_1 a$$

$$M_{m2} = R_3 b$$

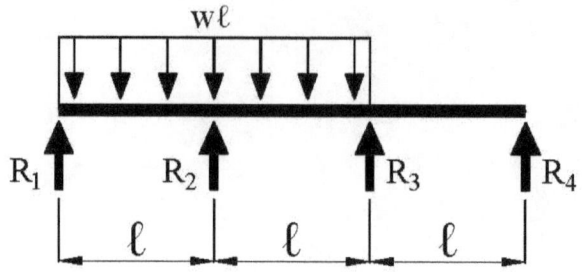

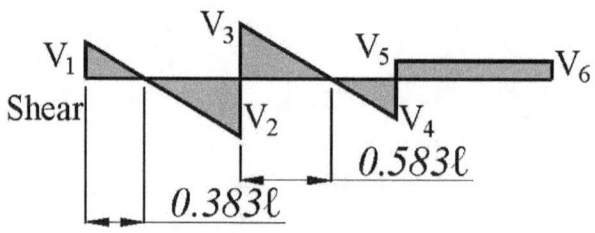

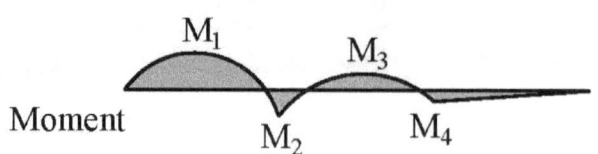

Continuous Beam

Three Span with Two Span Uniformly Distributed Load

$R_1 = V_1 = 0.383w\ell$

$R_2 = 1.2w\ell$

$R_3 = 0.45w\ell$

$R_4 = -0.033w\ell$

$V_2 = 0.617w\ell$

$V_3 = 0.583w\ell$

$V_4 = 0.417w\ell$

$V_5 = V_6 = 0.033w\ell$

$M_1 \ (at\ 0.383\ell\ from\ R_1) = 0.0735w\ell^2$

$M_2 \ (at\ R_2) = -0.117w\ell^2$

$M_3 \ (at\ 0.538\ell\ from\ R_2) = 0.0534w\ell^2$

$M_4 \ (at\ R_3) = -0.0333w\ell^2$

$\Delta_{Max} \ (at\ 0.43\ell\ from\ R_1) = \dfrac{0.0059w\ell^4}{EI}$

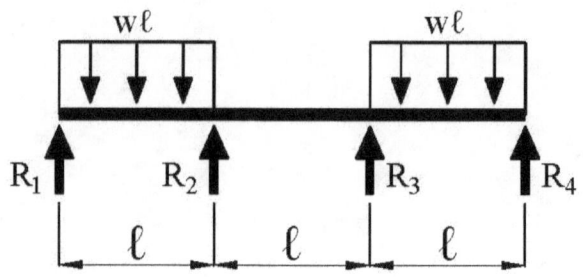

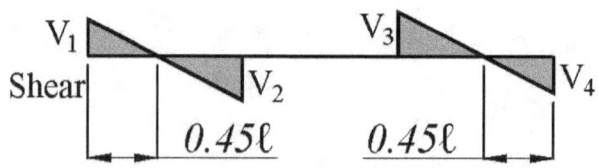

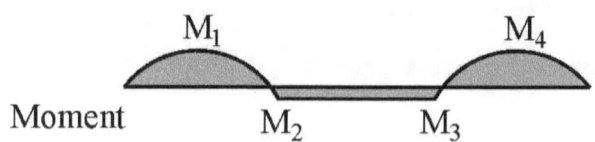

Continuous Beam

Three Span with End Span Uniformly Distributed Loads

$R_1 = V_1 = R_4 = V_4 = 0.45w\ell$

$R_2 = V_2 = R_3 = V_3 = 0.55w\ell$

$M_1 = M_4 \ (at\ 0.45\ell\ from\ R_1\ or\ R_4\ Respectively) = 0.1013w\ell^2$

$M_2 = M_3 \ (at\ mid\ span) = -0.05w\ell$

$\Delta_{Max} \ (at\ 0.479\ell\ from\ R_1\ or\ R_4) = \dfrac{0.0099w\ell^4}{EI}$

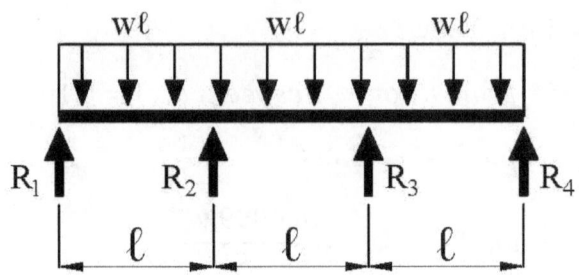
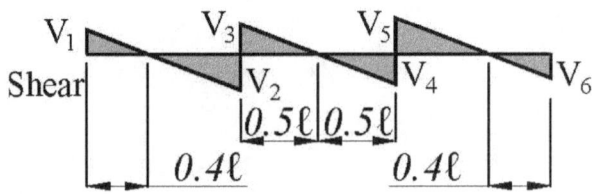
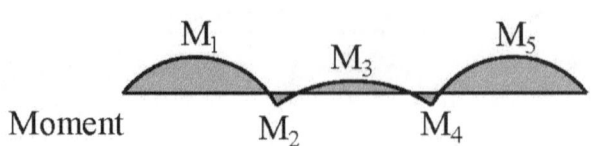

Continuous Beam

Three Span Beam with Uniformly Distributed Loads

$R_1 = V_1 = R_4 = V_6 = 0.4w\ell$

$R_2 = R_3 = 1.1w\ell$

$V_2 = V_5 = 0.6w\ell$

$V_3 = V_4 = 0.5w\ell$

$M_1 = M_5$ (at 0.4ℓ from R_1 or R_4 Respectively) $= 0.08w\ell^2$

$M_2 = M_4$ (at R_2 or R_3) $= 0.1w\ell^2$

M_2 (at mid centre span) $= 0.025w\ell^2$

Δ_{Max} (at 0.446ℓ from R_1 or R_4) $= \dfrac{0.0069w\ell^4}{EI}$

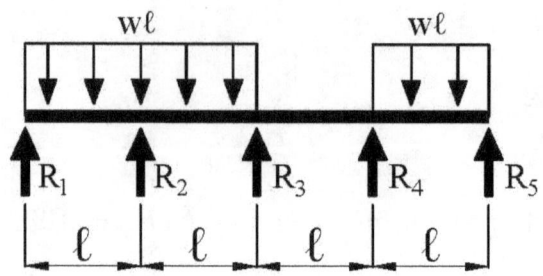

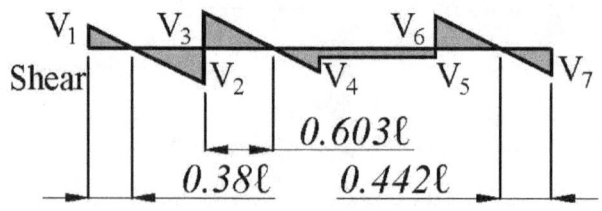

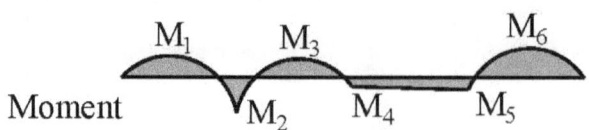

Continuous Beam

Four Span Beam with Unloaded Span

$R_1 = V_1 = 0.38w\ell$

$R_2 = 1.223w\ell$

$R_3 = 0.357w\ell$

$R_4 = 0.598w\ell$

$R_5 = V_5 = 0.442w\ell$

$V_2 = 0.62w\ell$

$V_3 = 0.603w\ell$

$V_4 = 0.397w\ell$

$V_5 = 0.04w\ell$

$V_6 = 0.558w\ell$

$M_1 \ (at\ 0.38\ell\ from\ R_1) = 0.072w\ell^2$

$M_2 \ (at\ R_2) = -0.1205w\ell^2$

$M_3 \ (at\ 0.603\ell\ from\ R_2) = 0.0611w\ell^2$

$M_4 \ (at\ R_3) = -0.0179w\ell^2$

$M_5 \ (at\ R_4) = -0.058w\ell^2$

$M_6 \ (at\ 0.442\ell\ from\ R_5) = 0.0977w\ell^2$

$\Delta_{Max} \ (at\ 0.475\ell\ from\ R_5) = \dfrac{0.0094w\ell^4}{EI}$

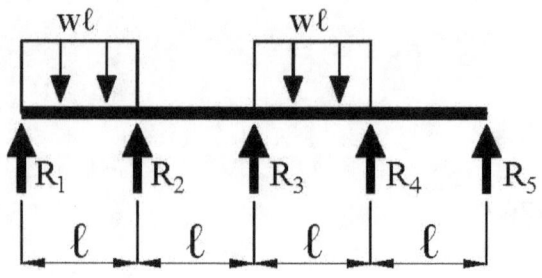

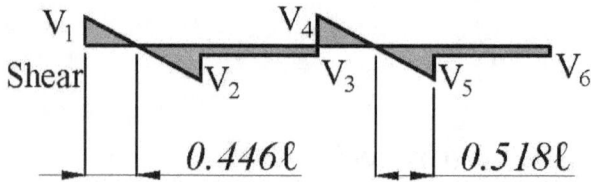

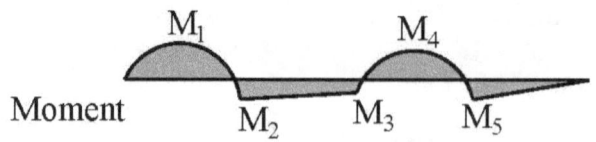

Continuous Beam

Four Span with Two Spans Unloaded

$R_1 = V_1 = 0.446w\ell$

$R_2 = 0.572w\ell$

$R_3 = 0.464w\ell$

$R_4 = 0.572w\ell$

$R_5 = -0.054w\ell$

$V_2 = 0.0554w\ell$

$V_3 = 0.018w\ell$

$V_4 = 0.482w\ell$

$V_5 = 0.518w\ell$

$V_6 = 0.054w\ell$

M_1 (at 0.446ℓ from R_1) $= 0.0996w\ell^2$

M_2 (at R_2) $= -0.0536w\ell^2$

M_3 (at R_3) $= -0.0357w\ell^2$

M_4 (at 0.518ℓ from R_4) $= 0.0805w\ell^2$

M_5 (at R_4) $= -0.0536w\ell^2$

Δ_{Max} (at 0.477ℓ from R_1) $= \dfrac{0.0097w\ell^4}{EI}$

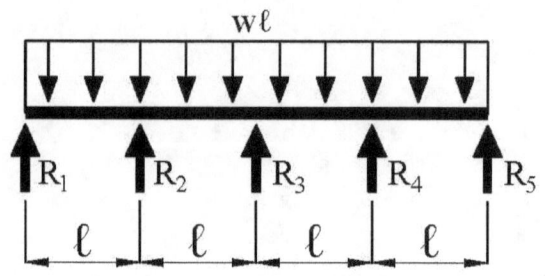

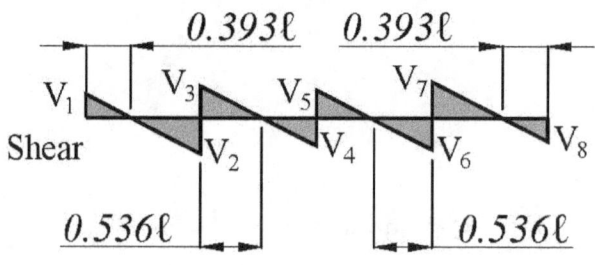

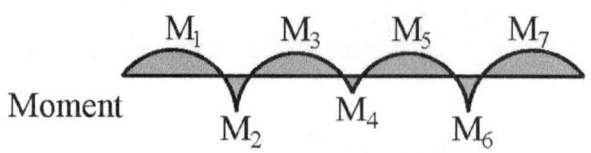

Continuous Beam

Four Span with UDL

$R_1 = V_1 = R_5 = V_8 = 0.393w\ell$

$R_2 = R_4 = 1.143w\ell$

$R_3 = 0.928w\ell$

$V_2 = V_7 = 0.607w\ell$

$V_3 = V_6 = 0.536w\ell$

$V_4 = V_5 = 0.464w\ell$

$M_1 = M_7$ (at 0.393ℓ from R_1 or R_5 Respectively) $= 0.0772w\ell^2$

M_2 (at R_2) $= -0.1071w\ell^2$

$M_3 = M_5$ (at 0.536ℓ from R_2 or R_4 Respectively) $= 0.0364w\ell^2$

M_4 (at R_3) $= -0.0714w\ell^2$

Δ_{Max} (at 0.44ℓ from R_1 & R_5) $= \dfrac{0.0065w\ell^4}{EI}$

www.ingramcontent.com/pod-product-compliance
Lightning Source LLC
Chambersburg PA
CBHW070115230526
45472CB00004B/1273